A VERY BERRY COOKBOOK

Compiled and Edited by
Judith Bosley

Cover Art
Jennifer L. Kuhn

Graphic Design
Ed Robson

D0393590

L.E.B. Inc.
Livonia, Michigan

First Printing March 1991
Printed in the United States of America
ISBN: 0-930809-10-6

L.E.B. Inc.
27599 Schoolcraft
Livonia, Michigan 48150

Take your Pick

- Strawberries
- Raspberries
- Blueberries
- Cranberries

There are at least two ways to pick your own strawberries. You can get up early on a warm June morning, take a straw hat because it will be hot later, and go to the strawberry patch. There you can squat, if you squat well, or get on your knees. Lift the leaves of the strawberry plants that grow right on the ground and capture your red prizes. There is a down-to-earth satisfaction in this down-to-earth exercise of food gathering that probably reaches back through all our generations. Three quarts is about the limit of my enjoying this exercise though. The second method is to pull off the highway at a summer fruit market and pick out enough fresh picked berries for the jams, shortcake and many other treats you will find between these pages.

Raspberries are a different, and more picky story. In Michigan and other northern states, hot, midday in July is the time to pick these choice morsels. Raspberries grow on stalks that can be shoulder-high. Only those willing to risk bloodshed for fruit, pick these. But Ohhh, the sweet treat of a warm red raspberry on the tongue. Pick one, eat one, is my motto, so I seldom come home with many for dinner. If you are fortunate enough to have access to the beautiful, tart black raspberries, they can be used in any recipe by increasing sugar. They can also be mixed with the red variety to add tartness, and they make superb jam. Pick raspberries when they are ready or the birds will pick them for you!

Although Michigan is tops world-wide in the production of blueberries, and although they grow on lovely little bushes, pretty enough to plant for shubbery, I've never picked a single blueberry; just huckleberries in the northwoods when I was a little girl. Since I associate little blue berries with being lost in the woods, I get all my blueberries in a nice safe supermarket, where I've seldom been lost. They are plentiful in season, keep up to two weeks when refrigerated, and freeze well. Freeze berries unwashed, then pour them out, wash, thaw and drain on paper towels, and use the same as fresh with the same results. If you are a blueberry lover, flip to that section and pick out a recipe to try. Some of them are from the private collection of a friend in Muskegon, Michigan who lives across the road from a blueberry field. Its almost like picking the berries yourself.

Cranberries are one of the three fruits native to America, the others being blueberries and Concord grapes. Probably the Pilgrims and Indians shared these crimson beauties on the first Thanksgiving, and possibly most Pilgrims and Indians are still grinding them with apples and serving them with turkey three hundred years later. You can't plan on picking your own cranberries, as this delicate morsel grows in bogs which are flooded at harvest time in a most interesting harvest operation. There is a museum just a short walk from Plymouth Rock that will show you the whole operation. Next to Massachusetts in cranberry production is Wisconsin, a longer walk from Plymouth Rock. Like bluberries, cranberries keep well when fresh and freeze well, which make them a good choice for flavoring winter salads, breads, hot drinks and desserts.

Contents

VERY RASPBERRY

VERY BLUEBERRY

VERY CRANBERRY

VERY STRAWBERRY

1   **Strawberry Banana Salad**
Two favorites combined, also good with raspberries

2 3 oz pkg strawberry gelatin   3-4 mashed bananas
2 C boiling water   1 C sour cream
2 10 oz pkgs frozen strawberries 1/2 C nutmeats
1 large can crushed pineapple with juice

Add water to gelatin and stir to dissolve. Stir in frozen berries,
pineapple and bananas. Pour half of mixture into an oiled mold or
9X13 pan; chill until set. Spread mixture with sour cream and
sprinkle with nuts. Add remaining fruit and chill until set.
Serves 12.

2   Strawberry Cream Cheese Squares
Serve with a sandwich for lunch

1 pkg strawberry gelatin
1 C boiling water
1 10 oz pkg frozen strawberries

1 small pkg whipped topping
1 3 oz pkg cream cheese
1 C miniature marshmallows

Dissolve gelatin in boiling water; add frozen berries and stir until thawed. Pour into a 9X9 inch pan and chill until set. Add softened cream cheese and marshmallows to whipped topping. Spread over gelatin. Cut in squares. Serves 9.

3 Strawberry Ice Cream Salad
Make in a minute for unexpected guests

1 pkg strawberry gelatin
1 C boiling water

1 10 oz pkg frozen strawberries
1 C vanilla ice cream

Dissolve gelatin in water; stir in berries and ice cream. Refrigerate for 20 minutes. Serves 4.

4 Strawberry Dressing
For fruit salad

1 C fresh strawberries
1/2 sugar
1/2 C currant jelly

1/3 C water
4 T cornstarch
2 T lemon juice

Combine berries, sugar, jelly and water; bring to a boil and simmer for 15 minutes. Mix cornstarch with 2 T water and stir into hot mixture. Cook one minute more and add lemon juice. Put hot mixture through a sieve. Chill. Serve dressing over mixed fruits such as peaches, pears, bananas, grapes, apples and oranges.

5   Ruby Red Salad
A pretty Christmas salad

2 3 oz pkg strawberry gelatin
2 C boiling water
1 #2 can crushed pineapple

2 C whole cranberry sauce
1 10 oz pkg frozen strawberries
1 3 oz pkg cream cheese

Dissolve gelatin in boiling water. Stir strawberries into hot liquid, then add cranberries and pineapple, which has been drained (reserve juice). Pour salad into an oiled mold or serving dish. Whip reserved pineapple juice into cream cheese; serve on the side with salad. Serves 6-8.

6   **Russian Strawberry Soup**
A different, delicious appetizer

4 C very ripe strawberries
1 C sugar
1 C sour cream
4 C cold water

Remove hulls, wash berries and press through a sieve. Add sugar,
sour cream and water; heat slowly, stirring constantly just until hot,
stirring constantly with a wooden spoon. Do not boil! Serve hot
or cold, with a dollop of sour cream.

7   Strawberry Wine

A favorite of wine-makers

7 lbs strawberries	juice of 1 lemon
2 gallons boiling water	5 lbs sugar

Mash cleaned berries in a crock; add water and lemon juice, and stir vigorously. Cover with a cloth and let stand for one week, stirring once each day. Strain through a cheesecloth into a large bowl and discard fruit. Return juice to crock and stir in sugar. Cover again with a cloth and let stand for one week, stirring each day. Transfer juice to gallon jugs and leave not more than a two inch air space at the top of the jug. Top jugs with loose corks (so that air can escape) or with fermentation locks. Let jugs rest undisturbed for two to three months, then rack the wine. Bottle when fermentation has stopped and wine is clear. Age for one year. About 2 1/2 gallons of wine.

❧ A layer of sediment forms in the bottom of the jugs. "Racking" is the process of *siphoning* the wine into another jug, without disturbing the sediment, and without the wine being exposed to the air as it would be if poured.

8   Strawberry Cooler
Vary according to taste and calories desired

1 C fresh or frozen strawberries
1 C yogurt, plain, vanilla or strawberry
1 C milk, whole skim or 2 %
2 T honey or sugar

Mix in a blender until smooth and serve. Serves 3.

❦ To make **a fizz**, substitute club soda for milk.❦

9 Strawberry Amaretto
Rich strawberry-almond dessert drink

1 pt fresh or frozen berries & juice   1 pt vanilla ice cream
3-6 T amaretto   12 ice cubes

Blend until smooth. Serves 6.

10 Strawberry Daquiries

1 pt frozen strawberries   6-12 ice cubes
1 6 oz can pink lemonade   6 oz rum

Blend until smooth. Serves 6.

(To make punch for a crowd, double all ingredients and add 1 2
liter bottle mix of choice.)

11   Strawberry-Apple Punch
Perfect punch for June graduation parties

2 qt strawberries   2 qt apple juice
1 2 liter bottle mix of choice   mint leaves

Blend berries and 2 C juice in a blender until smooth. Pour mixture into punchbowl and add remaining juice and mix. Float ice ring in bowl

❧ Make ice ring by freezing one inch of ice in a ring mold; add small amount of water to top of ice and arrange reserved berries with mint leaves on ice and freeze again. Fill mold to top with water and freeze again.

❧ Freeze a few large, perfect berries with stems on in a plastic bag. Keep on hand to drop in a glass of wine or a soft drink

12   No Cook Strawberry Jam
Try this with raspberries too

3 C crushed fruit   1 C water
5 C sugar   1 pkg pectin

Mix berries and sugar thoroughly. Bring water and pectin to a boil, and boil for 1 minute. Pour hot mixture into fruit and stir for 2 minutes. Put into sterilized jars and seal or freeze.

🍃 Variations: Add 1/4 t each cinnamon, allspice and cloves to fruit for spiced jam; add 1 T finely chopped crystallized ginger for ginger jam; add 1 T grated lemon or orange rind for citrus jam.

13 Old Fashioned Strawberry Jam
Like grandma made on the old cookstove

6 C strawberries
1/2 C lemon juice
6 C sugar

Wash, hull and leave berries whole. Cover berries with boiling
water and let stand for 3 minutes, then drain off water. Combine
berries and 3 cups sugar in kettle; bring to the boiling point and
boil for 8 minutes, stirring constantly. Add remaining sugar and
lemon juice; boil for 10 minutes more, stirring constantly. Remove
from heat; stir and skim for 2 minutes. Pour jam into a shallow
glass baking dish and cool completely. When cold, put in jars and
seal.

14   No-Sugar Strawberry Jam
Try this with other fruits too

1 1/2 C unsweetened frozen berries
1/4 C frozen apple juice concentrate
1 1/2 T tapioca

Mix ingredients and let stand for 5 minutes. Bring to a boil stirring constantly. Cool for 20 minutes. Put in jars and store in refrigerator.

15   Strawberry-Rhubarb Jam

2 3 oz pkg strawberry gelatin
5 C chopped rhubarb
5 C sugar

Mix rhubarb and sugar and let stand overnight. Bring to a boil and cook for 2 minutes; add dry gelatin and cook for 10 minutes, stirring to break up rhubarb. Pour into hot sterilized jars.and seal.

16 Christmas Strawberries
A sweet gift

4 C grated coconut (12 ozs)
1 15 oz can sweetened condensed milk⊷
2 3 oz pkg strawberry gelatin
1 T sugar
red sugar
artificial leaves and stems

Combine coconut, milk, and dry gelatin, and blend thoroughy.
When firm enough to handle, form into strawberry shapes. Roll in
red sugar, and insert stems and leaves. Allow to dry.

⊷ You can make your own substitute:

1 C non-fat dry milk   1/3 C boiling water
3 T melted butter   2/3 C sugar

Blend ingredients until smooth. This amount will substitute for
one can of sweetened condensed milk.

17   Cream Dip

1 beaten egg
1/2 C sugar
2 T lemon juice

2 t grated lemon peel
1 C whipping cream
1 T grated orange peel

Combine egg, sugar, juice and grated fruit peel; cook for 3 minutes and cool. Whip cream and fold in cooled mixture. Serve with fresh strawberries for dipping.

18   Pina Colada Dip

2 C pineapple yogurt (not swiss style)
1 8 oz carton frozen whipped topping, thawed
2 pkg Pina Colada mix
flaked coconut to taste

Mix together and refrigerate overnight. Serve in a hollowed out canteloupe or pineappple surrounded by strawberries and other fruit as desired.

19   Kuhlua Dip

This coffee liqueur gives rich flavor

1 8 oz pkg cream cheese
2 T brown sugar

1 C marshmallow cream
1 T kahlua or more to taste

Combine ingredients and mix well.

20   Strawberry Bourbon Fruit Sauce
Serve over ice cream

1 C brown sugar

1 C white sugar

1 C water

1 C broken pecans

1 C strawberry preserves

1 orange

1 lemon

1/2 C bourbon

Cook sugars and water to 240 degrees on a candy thermometer. Remove from heat and stir in pecans and preserves. Remove rind from orange and lemon with a potato peeler; chop rind finely. Remove white membrane from both fruits and cut fruit in small pieces. Add rind, fruit and bourbon to first mixture. Refrigerate for a few hours for flavors to develop. Keeps indefinitely.

21   Strawberry Wine Sauce

2 C strawberries   1/2 C red wine
1/4 C water   1/4 C sugar

Combine 1/2 of berries and rest of ingredients and cook until
thickened. (Can be done in microwave.) Strain mixture, pressing
solids through strainer. Stir in remaining berries. Refrigerate for 3
hours. Serve with pound cake or over fruit salads.

22   Strawberry Butter
Try this on toast and pancakes

3/4 C fresh strawberries
3/4 C (1 1/2 sticks) butter
1 T honey

Puree berries in a blender. Have all ingredients at room
temperature and gradually whisk berries and honey into butter.
Chill.

23 Strawberry Bread

Berries and cream in a new form

1 1/2 C mashed strawberries
1 1/2 C flour
1/2 t baking soda
1/2 t cinnamon
1 C sugar

1/2 t salt
1/2 C oil
2 eggs
3 oz cream cheese (for topping)
1/2 C nutmeats (optional)

Drain 1/4 C juice from berries and pour over cream cheese and let stand at room temperature. Mix flour, soda, cinnamon, sugar and salt in a mixing bowl. Make a hole in the center of mixture. Add berries, oil and eggs. Mix by hand until combined. Stir in nuts. Pour mixture into a greased loaf pan. Bake at 350 degrees, 45-55 minutes, or until bread tests done. Mix juice into softened cream cheese. Spread cheese mixture over top of warm loaf.

24   Fresh Strawberry Coffee Cake

1/2 C sugar
1 C flour
2 t baking powder
1/2 t salt
1/2 C milk
1 egg
2 T melted butter
2 C fresh strawberries, chopped

Topping:
1/2 C flour
1/2 C sugar
1/4 C butter
1/4 C chopped nuts

Combine all batter ingredients except berries, and mix just until blended. Spread mixture in a greased 8 inch square pan. Arrange berries evenly over batter. Combine topping ingredients and sprinkle over berries. Bake at 375 degrees, 35-40 minutes. Serve warm.

25 Strawberry Breakfast Cake
Your guests will rave about this!

2 T butter, melted
3 eggs, beaten
1 1/2 C milk
3/4 C flour
1/3 C sugar
1/4 t salt

3 C fresh strawberries, sliced
2-3 T sugar

1 C sour cream
1/4 C brown sugar

Preheat oven to 375 degrees. Toss strawberries with sugar and set aside. Pour melted butter in a 9 inch pie pan and swirl to coat side and bottom. In mixer bowl combine eggs and milk; add flour, 1/3 C sugar and salt. Beat until smooth. Pour batter into buttered pan and bake for 30 minutes or until edges are brown and center is set. (A hollow well will form in the center.) Remove from oven, spoon berries in the center and serve immediately. Cut in wedges and serve with a dollop of sour cream and a sprinkle of brown sugar. Serves 6.

26 Strawberry Shortcut Cake

1 pkg white cake mix   2/3 C oil
1 3 oz pkg strawberry gelatin   4 eggs
1/2 C hot water
1/2 C frozen strawberries, thawed

Mix gelatin powder with cake mix; add rest of ingredients and mix well. Bake in a 13X9 inch pan for 40-50 minutes. Cool completely. Spread topping on cake and garnish with berries.

❧ Topping: 1 pkg instant strawberry pudding mix; 1 C milk, 1 small pkg frozen whipped topping, thawed. Stir milk into pudding and let thicken. Stir in frozen topping. (This topping is good made with other kinds of puddings for other kinds of cakes.)

27   Strawberry Filled Angelfood Cake
Each slice has a strawberry center

1 angel food cake
1 3 oz pkg strawberry gelatin
1 C boiling water
1 10 oz pkg frozen strawberries, or use fresh berries
1 1 small carton frozen whipped topping, thawed

Cut top off cake about 1/3 down from top. With a cake knife and a spoon, gently remove center of bottom layer, reserving cake that is removed. Dissolve gelatin in boiling water and add frozen berries, reserving a few slices for garnish. Chill until almost set, then stir in reserved cake pieces. Fill center of cake with gelatin mixture, put top back on cake. Chill. At serving time, frost cake with topping and garnish with reserved berries. (If fresh berries are used, mash berries slightly and add 2 T sugar to 1 C berries.)

❧ *Strawberry Shortcake* ❧

This is possibly the favorite American dessert of them all. There are as many variations as there are families. Some like one large shortcake split horizontally, buttered, spread with sliced berries in the middle and on top, and cut in wedges. Some like small biscuits with mashed, sweetened berries topped with ice cream. Our family favorite is the puffy #2 shortcake, hot from the oven, heaped with finely chopped, sweetened berries and a dribble of milk.

28 Old-Fashioned Shortcake
Brown sugar is a nice touch

2 C flour
3 t baking powder
3/4 t salt

1/4 C brown sugar
1/2 C butter or margarine
1/2 C milk (may need more)

Combine dry ingredients and cut in butter with a pastry blender. Make a well in center and add milk all at once. Stir with a fork just until dough cleans the side of the bowl; use slightly more milk if needed. Turn dough onto a lightly floured board and knead 10 times. Pat dough into a greased layer cake pan, or cut small biscuits. Bake at 450 degrees, 20 minutes or until nicely browned. Serve hot.

29 Shortcake #2
Quick and very tender

2 C flour
3 T sugar
3 t baking powder
1 t salt

1/3 C oil
2/3 C milk

Mix oil and milk. Combine dry ingredients and add liquid all at once. Stir with a fork until mixture forms a ball. Turn onto a floured board; pat out 3/4 inch thick and cut in rounds with a floured glass or cookie cutter. Bake on an ungreased cookie sheet at 475 degrees, 10-12 minutes.

30 Shortcake #3
Make one large or individual shortcakes

2 C flour
1/2 C sugar
4 t baking powder

1/2 t salt
1 1/4 C half and half

Combine dry ingredients and stir in half and half with a fork just until blended. Cut biscuits or pat dough into a greased, round cake pan. Bake at 450 degrees, 12-15 minutes.

31   Frozen Strawberry Squares
A cool treat for a hot day

Crust:
1 C flour
1/2 C brown sugar
1/2 C nuts
1/2 C butter
2 T lemon juice

Filling:
2 egg whites
3/4 C sugar
1 pkg frozen berries or
2 C fresh berries

1 C whipping cream, whipped, or 1 small carton whipped topping

Mix crust ingredients and bake in a 13x9 inch pan at 350 degrees
for 20 minutes, stirring occasionally. Cool and reserve one-third
of mixture, spreading remaining crumbs evenly over bottom of
pan. Combine egg whites, sugar, berries and lemon juice in large
mixer bowl and beat until blended, then turning on high, beat
10-12 minutes or until stiff peaks form. Fold in whipped cream or
topping. Pour over crumb crust and sprinkle remaining crumbs on
top. Freeze. Cut in squares. Serves 12.

32   Strawberry Angel Dessert
Angel food and berries are heavenly

1 angel food cake
1 can sweetened condensed milk
1 qt fresh strawberries, mashed
4 T lemon juice
1 12 oz carton whipped topping

Break cake into walnut-sized pieces in a 13x9 cake pan (preferably glass). Combine berries, milk and lemon juice. Spoon mixture over cake. Spread whipped topping over berries. Chill for several hours. Garnish with whole berries.

33 Strawberry Wonder

If you are going to sin, try this!

1 lg box vanilla wafers, crushed
1 stick butter (no substitute)
1 lb powdered sugar
2 eggs
1 pt whipping cream
2 qts fresh strawberries

Sprinkle half of crumbs in a 13x9 pan (preferably glass). Cream softened butter and sugar; beat in eggs one at a time and continue beating until very creamy. Spread mixture carefully over crumbs. Wash and hull berries; layer berries over butter mixture. Whip cream until quite stiff and spread over berries. Top with remaining crumbs. Cover and refrigerate 8 hours or overnight. Top each serving with an unstemmed berry.

34   Homemade Strawberry Ice Cream

5 eggs   2 C sugar
1 qt half and half   1 can evaporated milk
1/2 t salt   1/3 C flour
3 C chopped or mashed strawberries, sweetened
2 C milk (approximately)

Beat eggs and sugar until light and creamy; add evaporated milk, salt and flour and mix thoroughly. Stir berries into mixture and put in freezer. Fill freezer with milk. Freeze according to freezer directions. 1 gallon.

35   Strawberry Sherbert
So pink and so pretty; not as rich as ice cream

1 qt milk
2 qts strawberries
2 C sugar

Mash berries and drain off juice; (save for another recipe) Mix berries with sugar and milk. Freeze according to ice cream freezer directions. Makes 3 quarts sherbert.

36 Strawberry-Rhubarb Crunch
Good complimentary flavors

1 C flour
3/4 C oatmeal
1 C brown sugar
1/2 C butter, melted
2 1/2 C rhubarb
1 t cinnamon

1 1/2 C strawberries
1 C sugar
2 T cornstarch
1 C water
1 t vanilla

Mix oatmeal, flour, brown sugar, butter and cinnamon until crumbly. Press half of mixture into a greased a 9X9 inch pan. Cover with strawberries that have been sliced, and diced rhubarb. Combine sugar and cornstarch; stir in water and cook until thick and clear. Add vanilla and pour mixture over fruit. Top with remaining crumbs. Bake at 350 degrees, 50-60 minutes.
Serves 6-8

37 Strawberry Glazed Cheesecake
Picture pretty and not a bit hard to make

1 1/3 C graham cracker crumbs
1/4 C sugar
1/4 C butter, melted
1 lb creamed cottage cheese
2/3 C sugar
2 T flour

3 eggs
1 C evaporated milk
1/2 t salt
1/2 t vanilla
2 T lemon juice
10-12 large strawberries

Combine cracker crumbs, 1/4 C sugar and butter. Press mixture on bottom and sides of an 8X8 in baking pan or 8 inch springform pan. Chill. Beat cottage cheese at high speed until smooth; add sugar and flour and continue to beat. Add eggs one at a time; blend in milk, salt, lemon juice and vanilla. Pour mixture over crumbs. Bake at 350 degrees, 50 minutes or until set. Cool. Place berries, pointed end up around edge of cheesecake. Glaze.

Strawberry Glaze:
1 pt fresh berries
1 1/2 T cornstarch

1/2 C sugar
red food color

Crush berries; mix with sugar and cornstarch. Cook until thick and clear. Add food coloring if desired. Spread glaze over top of cake including berries.

38 Strawberry Yogurt Cream Pie
This is good with many kinds of fruit

1 8 inch pie shell, baked
1 C plain yogurt
1 8 oz pkg cream cheese

3 T honey (or to taste)
1/2 t vanilla
1 C sliced strawberries

Combine yogurt, 1 T honey, cream cheese and vanilla. Beat until light. Pour into pieshell and refrigerate. At serving time, sweeten berries to taste with honey and spoon on top of filling. Serves 6.

39 Strawberry Ice Cream Pie

1 pkg lemon gelatin
1 pt vanilla ice cream
1 pie shell or graham cracker crust

1 C hot water
1 C fresh or frozen strawberries

Dissolve gelatin in water; stir in ice cream immediately. Chill for 10 minutes. Stir in strawberries and pour into pie shell. Chill or freeze.

40 Strawberry Alaska

Easy to make, fun to serve

1 9 in baked pie shell
2 T sugar
1/4 t cream of tartar
1 pt lemon sherbert or strawberry ice cream

1 pt strawberries
3 egg whites
3 T sugar

Spoon sherbert or ice cream into pie shell and spread gently . Cover and freeze until firm. Sweeten berries with 2 T sugar and chill. At serving time, beat egg whites and cream of tartar to soft peaks; gradually add 3 tablespoons sugar, beating until stiff peaks form. Spoon berries over sherbert. Spread meringue mixture over berries to edge of pie, sealing it to edge of pastry to prevent shrinking. Bake at 500 degrees for 2-3 minutes or until golden. Watch carefully! Serves 6-8.

41 Strawberry Rhubarb Pie

1 1/4 C sugar
1/4 t salt
1/3 C flour
Pastry for a two crust pie

2 C diced rhubarb
2 C fresh strawberries
2 T butter

Combine sugar, salt and flour. Arrange half of fruit in pie shell,
sprinkle with half of sugar mixture; repeat with remaining fruit and
sugar. Dot with butter. Put on full top crust or lattice crust.
Sprinkle top crust with sugar. Bake at 400 degrees, 40-50 minutes
or until rhubarb is done and crust is browned.

42 Glazed Strawberry Pie

Our all-time favorite. The touch of green in the garnish makes it as beautiful as it is delicious.

1 qt ripe strawberries
3 T corn starch
1/2 C water
1 9 inch baked pie shell
Whipped cream or frozen whipped topping

1 T lemon juice
1/4 t salt
1 C sugar

Cut half of berries in half, reserving 6 small berries with stems left on for garnish. Put halved berries in pie shell. Crush remaining berries and combine with sugar, water, cornstarch, lemon juice and salt; cook until thick and clear. Cool. Pour cooked mixture over berries in pie shell. Chill. Put a ring of whipped cream around top of pie and decorate with reserved berries.

43   Strawberry Pizza

Crust
1/2 C powdered sugar   1 1/4 C flour
1/2 C butter

Combine ingredients and press onto a pizza pan, forming a rim.
Bake at 350 degrees, 15 minutes. Cool.

Base
1 8 oz pkg cream cheese   1 t vanilla
2/3 C powdered sugar

Combine and spread over cooled crust.

Glaze
1 C strawberry juice   1 T cornstarch
1/3 -1/2 C sugar

Cook until thick and cool. Add 3 C sliced berries to glaze and
spread over cheese layer. OR, place fresh berries in a design over
cheese and pour glaze over fruit.

⁋ This new dessert can be made with other fruits and juices, and
with fruits in combination; pineapple, peaches, cherries, grapes,
and kiwi fruit.

VERY
RASPBERRY

44   **Raspberry Pretzel Salad**
Also good with strawberries

1 C crushed prezels topping
1 1/2 T sugar
4 T margarine
1 3 oz pkg cream cheese
1/2 C sugar

1 small carton frozen whipped

3 oz pkg raspberry gelatin
1 C boiling water
1 10 oz box frozen raspberries

Mix pretzels with 1 1/2 T sugar and soft margarine. Press mixture into an 8X8 inch pan. Bake at 350 degrees, 10 minutes. Cool. Combine sugar and softened cream cheese, and fold in thawed topping. Spread over crust. Dissolve gelatin in water and add frozen berries stirring to break up berries. Pour berries over topping and chill until set. Recipe may be doubled.

45   Raspberry Crown Salad

So pretty for a holiday dinner.

1 3 oz pkg lemon gelatin
1 C boiling water
1 8 oz pkg cream cheese
1 C half and half
3 T powdered sugar
1 t vanilla

2 3 oz pkg raspberry gelatin
2 1/2 C boiling water
2 10 oz pkgs frozen raspberries,
  thawed and drained
1 C raspberry juice

Dissolve lemon gelatin in water; add softened cream cheese, half and half, sugar and vanilla, beating until smooth. Pour mixture into an oiled mold and chill until set. Dissolve raspberry gelatin in water; add berries and juice. Chill until partially set, and stir to prevent floating fruit. Pour on top of first mixture and chill again until set.

46   Red Raspberry Ring

1 10 oz pkg frozen raspberries
2 3 0z pkgs raspberry gelatin
2 C boiling water

1 pt vanilla ice cream
1 6 oz can frozen pink lemonade
1/4 C chopped pecans

Thaw and drain berries, reserving syrup. Dissolve gelatin in water; add ice cream, stirring until melted. Stir in lemonade concentrate and reserved syrup. Chill until partially set; fold in berries and nuts. Pour into an oiled ring mold. Chill until set.

47   Raspberry Applesauce Salad

2 pkgs raspberry gelatin
1 1/2 C boiling water

1 10 oz pkg frozen rasperries
2 C applesauce

Dissolve gelatin in water; add frozen berries and stir until thawed. Chill until thickened, but not set; add applesauce. Pour into an oiled mold and chill until set.

48 Raspberry Gelatin Molds
Can make in four individual molds

1 3 oz pkg raspberry gelatin
1 C boiling water
1 1/2 C applesauce
1/3-1/2 C salad dressing

1 C drained pineapple tidbits
1 C miniature marshmallows
1/3 C chopped walnuts

Dissolve gelatin in water; chill until slightly thickened. Add applesauce and pour into a square pan or individual ring molds. Mix marshmallows, pineapple and nuts with enough salad dressing just to moisten. Serve on top of molded squares or in center of ring molds.

49 Raspberry Dressing
Invented for a no salt diet. Delicious!

1/2 C bottled raspberry syrup
1/2 C oil
1/4 C cider vinegar

Shake together. Do not cook.

Tear very crisp lettuce into a bowl. Top with crisp celery slices
and broken walnuts. Toss with dressing.

50   Raspberry Punch

3 3 oz pkgs raspberry gelatin
4 C boiling water
1 1/2 C sugar
4 C cold water
1/2 C lime juice

2 1/4 C orange juice
1 1/4 C lemon juice
1 qt ginger ale
2 10 oz pkgs frozen raspberries

Dissolve gelatin in boiling water; add sugar, cold water and juices. At serving time, place frozen berries in punch bowl; pour in punch and ginger ale.

❦ Raspberry Cubes: Fill an ice cube tray with water. Drop in a fresh raspberry and freeze. Serve in lemonade

51 Raspberry Cordial

2 qts fresh raspberries
2 C water

2 C sugar
2 fifths whiskey

Mix sugar and water until sugar is dissolved. Pour liquid into a gallon jug. Add berries and whiskey. Cover jug and let mixture stand 2-4 months. Strain through a cheese cloth and bottle.

52 Raspberry Liqueur

2 10 oz pkgs frozen raspberries
1 1/2 C sugar
1 1/2 C vodka

Thaw berries in refrigerator; drain off juice. Add sugar to juice and boil, stirring constantly for 5 minutes. Remove from heat and cool for one hour, skimming off any collected foam. Stir berries and vodka into liquid. Pour mixture into a decanter or jar. Cap and let stand in a dark place for one month, shaking bottle occasionally to mix. Strain through cheese cloth to serve. Makes 3 cups.

53 Raspberry Jam
Straight from a berry grower

5 C raspberries
4 C sugar
1 6 oz pkg raspberry gelatin

Combine berries and sugar and cook for ten minutes, stirring as
needed. Remove from heat and stir in gelatin. Pour into prepared
jars and seal or freeze.

꙾ See also strawberry jam recipes. Raspberries may be
substituted.

54   Raspberry Jam Bars

2 1/4 C flour
1 C sugar
1 C chopped pecans

1 C butter or margarine
1 egg
10 oz raspberry preserves

Mix all ingredients except preserves until crumbly. Reserve 1 1/2 C of mixture and press the remaining crumbs in a 13X9 pan. Spread preserves to within 1/2 inch of edge of crust. Sprinkle remaining crumbs over preserves. Bake at 350 degrees, 40-50 minutes or until brown.

❧ Make your own fruity yogurt: Fold 1/2 C unsweetened berries into 1 C plain low-fat yogurt; add two packets artificial sweetener.

55 Raspberry Chews

Not too sweet, just right

1 10 oz pkg frozen raspberries
4 t cornstarch
1 pkg yellow cake mix
1/3 C quick oats

1/2 C margarine
1 egg
1/4 C quick oats
1 T margarine

Thaw berries and blend with cornstarch; cook until thick and clear.
Combine cake mix with 1/3 C oats and 1/2 C margarine until
crumbly. Reserve one cup of crumbs for topping and add egg to
the remaining crumbs and mix well. Press mixture in bottom of a
greased 9X13 pan. Bake at 375 degrees, 10 minutes. Spoon berry
mixture over baked crust. Mix remaining crumbs with 1/4 C oats
and 1 T margarine; sprinkle over berries. Bake 15-20 minutes
longer or until golden. Cool and cut into squares.

56 Raspberry Bars
Nuts and berries

Crust:
1 1/2 C flour
1 1/2 C quick oatmeal
1 C coconut
1 C slivered almonds
1 C brown sugar
1/2 C soft butter
2 T water

Filling:
2 10 oz pkgs frozen raspberries
2 1/2 T cornstarch
1/2 t lemon juice

Combine all crust ingredients and mix well. Press half of mixture in a 13X9 inch baking pan. Combine filling ingredients in a sauce pan and cook until thick and clear; cool. Pour berries over crust and put remaining crust mixture on top. Bake at 350 degrees for 35-40 minutes or until light brown. Cool and cut into bars.

57 Raspberry Jelly Roll
This is an old-fashioned dessert

3 eggs, separated
1 C sugar
1 C flour
1 t baking powder

1/4 t salt
1/4 C hot milk
1 t vanilla
2 C raspberry jam

Carefully grease and flour a jellyroll pan (cookie sheet with sides) or spray pan thoroughly with cooking spray. Prepare a towel somewhat larger than the pan by spreading it out and sprinkling it well with powdered sugar. Beat egg whites. Beat yolks separately and then blend together and beat again. Add sugar gradually while beating; add dry ingredients. Add milk and vanilla. Spread batter on prepared pan. Bake at 400 degrees, 10-12 minutes. Do not overbake. Invert hot cake immediately on towel. Roll cake, from the narrow end, rolling towel right into the roll of cake. Let stand until cool. Unroll and spread with raspberry jam. Reroll cake. Slice in 1 inch slices. (strawberry Jam and lemon filling are also good).

58 Apple Raspberry Crisp

4 large apples	2 C fresh raspberries
3 t lemon juice	1 t grated lemon rind
1 C flour	1/2 t salt
3/4 C brown sugar	5 T butter

Peel and slice apples into a 9X9 inch baking pan. Spread berries over apples; sprinkle fruits with lemon juice and grated rind. Combine flour, brown sugar, salt and butter. Stir half of flour mixture into fruit; pour remaining flour on top. Bake at 400 degrees, 35-40 minutes or until apples are done.
Serve with cream or ice cream.

59   Raspberry Blintz Dessert

Batter:
1/4 lb butter
1/2 C sugar
2 eggs
1 1/4 C flour
1 t baking powder
1/2 t salt
3/4 C milk

Filling:
1 lb ricotta cheese
2 T butter, melted
1 egg
2 t sugar
1/4 t salt

For batter: Cream butter and sugar, add eggs. Add dry ingredients alternately with milk and mix well. Combine filling ingredients. Grease a 9X9 inch baking pan and alternate layers of batter and filling. Bake at 350 degrees, 1 hour. Serve with Raspberry Sauce and a dollop of sour cream.

Blintz Raspberry Sauce: 2 10 oz pkgs frozen berries, 1 C red raspberry jam, 1 T kirsch or cherry brandy. Thaw berries. Put ingredients in a blender to puree. Strain mixture and discard seeds.

60   Creeping Crust Cobbler
Won't last long enough to get cold

1/2 C margarine
1 C flour
1 C sugar
1 t baking powder

1/2 C milk
2 C fresh raspberries
2 T sugar

Melt butter in 10 inch baking dish in oven. Mix flour, 1 cup sugar and baking powder; stir in milk and mix until blended. Spoon batter over melted butter. Heat berries with 2 tablespoons sugar just until hot. Pour over batter in baking dish. Bake at 350 degrees, 30 minutes or until crust is golden. Serve warm.

61   Raspberry Topped Cheese Cake

Crust:
 24 vanilla wafers
5 T melted margarine
Filling:
1 8 0z pkg cream cheese
1/2 C sugar
2 T lemon juice
1/2 t vanilla

1/8 t salt
2 eggs
1 C sour cream
2 T sugar
1/2 t vanilla

Combine vanilla wafer crumbs and melted margarine. Press into a 9 inch spring form pan. Chill. Beat softened cream cheese until fluffy; add 1/2 cup sugar, vanilla and salt. Add eggs one at a time, beating mixture well. Pour into crust. Bake at 325 degrees, 35-30 minutes. Remove from oven and top hot cake with sour cream, 2 tablespoons sugar and vanilla mixed together. Bake 10 minutes more. Cool and cut in narrow wedges. Spoon raspberry sauce over each piece.

Raspberry Cheesecake Sauce: 1 10 oz pkg frozen berries, 1/2 C sugar, 1/2 C currant jelly and 1 1/2 t cornstarch. Thaw berries and combine ingredients in a saucepan. Cool, stirring constantly, until thickened. Cool.

62   Danish Raspberry Pudding

2 1/2 C fresh raspberries   1 1/2 T flour
2 t lemon juice   1/4 t salt
1/4 C honey (or more to taste)   1 1/2 C milk
2 T butter   3/4 C plain yogurt

Mix berries with lemon juice and chill. Melt butter in a saucepan;
add flour, honey and salt. Add milk and cook, stirring constantly
until thickened. Cool and refrigerate. At serving time, stir yogurt
into berries and stir berry mixture into pudding. Serves 6-8.

63 Frozen Raspberry Swirl
Almost too pretty to eat

3/4 C graham cracker crumbs
3 T butter, melted
2 T sugar
3 eggs, separated
1 8 oz pkg cream cheese

1 C sugar
1/8 t salt
1 C heavy cream, whipped
1 10 oz pkg frozen raspberries

Combine crumbs, butter and 2 T sugar. Press mixture in 7X11 baking dish and bake at 350 degrees 6-10 minutes. Cool. Beat egg whites until stiff. Whip cream. Beat egg yolks until thick; add softened cream cheese, sugar and salt. Fold in egg whites and cream. Crush berries to a pulp in blender; swirl half of berries into creamed mixture and pour into crust. Put remaining berries on top and swirl with a knife. Freeze uncovered so that design will not be disturbed. Cover and return to freezer. Allow to stand at room temperature 20 minutes before cutting. Serves 8.

64 Raspberry Marshmallow Dessert
Delish!

2 10 pkg frozen raspberries
1 C water
1/2 C sugar
whipped cream)
2 t lemon juice
4 T cornstarch
1/4 C cold water

50 large marshmallows
1 C milk
2 C frozen whipped topping (or

1 1/4 C graham cracker crumbs
1/4 C chopped nuts
1/4 C butter

Heat berries with water, sugar and lemon juice. Combine water and cornstarch and stir into berries. Cook until clear and thickened. Cool. Melt marshmallows in milk over boiling water, or in microwave. Cool. Mix cream into marshmallow mixture. Combine crackers, nuts and butter. Press mixture into bottom of 13X9 inch pan. Spread marshmallow mixture in pan. Spread berries on top. Chill. Serves 12.

65 Fresh Raspberry Pie
Warm with ice cream

1 qt fresh raspberries
1- 1 1/2 C sugar (depending on sweetness of berries)
2 T tapioca
1 T flour
1/4 t salt
1 T butter
pastry for a two-crust pie

Lightly mix sugar, flour, salt and tapioca with washed, drained berries. Pour into an unbaked pieshell; dot with butter. Put on a decorative top crust. Sprinkle crust with sugar. Bake at 375 degrees, 1 hour. Cool before cutting.

66 Cherry-Raspberry Pie
A winning combination

2 C red raspberries
1 C pitted cherries
3 T tapioca
1 1/2 C sugar
1 t lemon juice
1 T butter
 pastry for a two crust pie

Mix fruits, add tapioca with sugar and lemon juice. Let stand one hour. Pour filling into pastry lined pan. Dot with butter. Brush edge of bottom crust with water, and put on top crust. Bake at 400 degrees, 10 minutes, 350 degrees, 30 minutes.

67 Mile High Raspberry Pie
A prize winner!

Crust:
1/2 C margarine
1/4 C brown sugar
1 C flour
1/2 C chopped nutmeats

Filling:
1 10 0z pkg frozen berries
1 C sugar
3 egg whites
1 C whipping cream

Combine crust ingredients until blended; spread on a cookie sheet and bake at 400 degrees, 12-15 minutes. Cool and spread crumbs in a 9-10 inch pie pan. A baked pieshell may also be used. Whip cream (or may substitute 8 oz. carton of whipped topping.) Beat thawed berries, sugar and egg whites in large mixer bowl for 15 minutes at high speed. Mixture will be like meringue. Fold in cream or topping. Pile mixture into prepared pie pan and freeze for 6-8 hours. Serve frozen wedges garnished with fresh berries.

68 Raspberry Dream Pie
No seeds. Just fruit and cream.

6 C raspberries
1/2 C water
2/3 C sugar

3 T corn starch
1 9 inch baked pie shell
1 C whipped cream or topping

Mash berries, heat and press through a sieve. Return berries to saucepan; add water and sugar mixed with cornstarch. Cook, stirring constantly until thickened. Cool. Pour into crust, top with cream, garnish with fresh berries. Chill.

69 Raspberry Glace Pie
The taste of fresh berries

1 qt raspberries
1 C water
1 C sugar
1 1/2 t cornstarch

1 T raspberry soft drink mix
1T unflavored gelatin
1/4 C cold water
1 graham cracker crust

Wash berries and drain. Cook water, sugar, corn starch and drink mix until clear. Add gelatin which has been softened in cold water. When mixture is cool and thickened, fold in berries and pour into prepared crust. Refrigerate until serving time. Serve with whipped cream.

VERY
BLUEBERRY

70 Blueberries and Snow Salad
White on top of the blue. Very pretty

1 3 oz pkg lemon gelatin
1 6 oz pkg lemon gelatin
1 C boiling water
2 14 1/2 oz cans blueberries, drained

1/4 C milk
1/2 pt whipping cream
1 3 oz pkg cream cheese

Dissolve small package of gelatin in boiling water. Beat softened cream cheese with milk or cream until smooth; stir into gelatin and chill until it begins to thicken. Whip cream and fold into mixture, (or substitute frozen whipped topping, thawed.) Pour mixture into a six-cup oiled mold and chill until firm. Add enough water to juice drained from berries to make 1 1/2 cups; heat to boiling and dissolve large package gelatin in juice. Chill until mixture is partially thickened. Fold in blueberries and pour over cheese mixture. Chill until firm. Invert on serving plate. Serves 12.

71   Mixed Up Fruit Salad
Try with either of these super dressings

2 C fresh blueberries
3 C melon balls, cantaloupe or watermelon
2 avocados, peeled and diced
1 T lemon juice

Prepare fruits ahead, sprinkle avocado with lemon juice. Pile into dessert dishes and drizzle with dressing

72   Sour Cream Dressing
A hint of mint

1 C sour cream   1 T lime juice
1 T sugar   2 T finely chopped fresh mint
(1 t dried mint may be substituted)

Combine ingredients and let stand for flavors to blend.

73 Honey Cream Dressing
Use as a fruit dip, or on any fruit salad

1/2 C honey
4 egg yolks, beaten
juice of one lemon
1/3 C olive oil

1/2 t salt
1/4 t paprika
1 C whipping cream

Heat honey until hot and runny; remove from heat and rapidly beat in egg yolks, all at once. Return to heat and cook for about three minutes, stirring constantly with a wire whisk until mixture is thickened. Cool slightly and beat in lemon juice, olive oil, salt and paprika. Whip cream until stiff, and fold honey mixture into cream until blended. Refrigerate until serving time.

74 Blueberry Salad Squares

2 3 oz pkgs blackberry gelatin
1/2 C sugar
1 15 oz can blueberries
1 small can crushed pineapple

1 8 oz pkg cream cheese
1 C sour cream
1/2 C chopped nuts

Dissolve gelatin in boiling water. Measure juices from berries and pineapple; add water to make 1 cup liquid and add to gelatin. Stir in fruit and chill in a flat dish until firm. Beat sour cream, cream cheese and sugar until fluffy; spread over congealed fruit. Sprinkle nuts on top. Cut in squares to serve. Serves 8-10.

75   **Purple Lady Salad**
Real whipped cream needed here

1 6 oz pkg raspberry gelatin
1 C boiling water
1 #2 can blueberries with juice
1 small can crushed pineapple with juice
1/2 C pecans, finely chopped
1/2 C whipping cream, whipped

Dissolve gelatin in water; add berries and pineapple with juices. Refrigerate mixture until almost set. Fold in pecans and whipped cream. Pile into a clear glass bowl or set in a flat pan and cut in squares. Serves 6.

76   Spiced Blueberry Jam

1 1/2 qts blueberries (to make 4 C prepared fruit)
2 T lemon juice
1/4 t cinnamon
1/4 t cloves
1/4 t allspice
4 C sugar
1 box powdered fruit pectin

Thoroughly crush berries (one layer at a time). Measure 4 cups
berries into a large sauce pan; add spices and pectin. Bring
mixture to a full boil. Immediately add all of sugar and bring to a
full, rolling boil; boil hard for one minute, stirring constantly.
Remove from heat and skim off foam.* Ladle into jars. Seal or
freeze.

❦ Hint: Stir 2 tablespoons butter or margarine into hot jam and
foam will disappear.

77 French Toast with Blueberries

1 loaf unsliced bread   1 1/4 C milk
1 pkg vanilla pudding mix   1/2 t vanilla
3 eggs   1 can blueberry pie filling

Slice bread one inch thick. Beat eggs until thick; add milk,
pudding mix (not instant) and vanilla. Dip bread into batter,
turning to coat each side. Fry on buttered skillet until golden.
Serve with butter and warmed fruit filling.

78 Blueberry Vodka Sauce

1 C blueberries   1/4 C vodka
3/4 C water   1/8 t nutmeg
1/2 C sugar

Cook berries, water and sugar until berries are cooked through.
Puree in a blender and then strain into a bowl. Stir in vodka and
nutmeg. Refrigerate for 3 hours. Serve over peaches or other fruit.

79   Blueberry Pancakes /Waffles⁓

2 eggs
1 C milk
1/4 C oil
2 C flour

3 t baking powder
1 t salt
1 T sugar
1 C blueberries

Beat eggs well; add milk and oil. Combine dry ingredients and add to liquids stirring just until blended. Fold in fresh berries, or frozen berries that have been thawed and drained on paper towels. Bake pancakes on a hot griddle.

⁓ To make waffles: Separate egg yolks from whites. Add yolks to batter as above. Beat egg whites until stiff and fold gently into batter with the blueberries. Bake in a waffle iron.

80 Blueberry Muffins
Is there any other kind?

2 1/2 C flour
2 t baking powder
1/2 t baking soda
3/4 C sugar
1/2 t salt

1 C buttermilk
2 eggs
1/2 C oil
1 1/2 C blueberries

Mix dry ingredients; stir in eggs, buttermilk and oil and stir just until blended. Gently fold in berries. (If using frozen berries, thaw and drain berries on paper towels before adding.) Fill well-greased or paper lined muffin cups 2/3 full. Bake at 400 degrees, 15-20 minutes. 12-16 muffins.

81 Blueberry Bran Muffins

1 1/2 C whole bran cereal
1 C buttermilk
1 egg, beaten
1/4 C oil
1 C flour

1/3 C brown sugar
2 t baking powder
1/2 t baking soda
1/2 t salt
1 C blueberries

Combine bran and milk and let stand for five minutes. Mix dry
ingredients. Blend sugar, egg and oil into bran; add dry
ingredients and stir just until flour is dampened. Fold in berries.
Fill prepared muffin cups 2/3 full. Bake at 400 degrees, 20-25
minutes. 12 muffins

82 Orange-Berry Muffins
An unusual new flavor

3 T sugar
1/4 C shortening
2 eggs
2 C flour
2 1/2 t baking powder

1/2 t salt
1 T grated orange rind
1/2 C milk
1/2 C orange juice
1 C fresh or frozen berries

Cream shortening, sugar and eggs; add dry ingredients, to which rind has been added, alternately with milk and juice. Fold in berries. Spoon mixture into prepared muffin cups and sprinkle each muffin with granulated sugar. Bake at 375 degrees, 15-20 minutes. 12 muffins.

83   Blueberry Whole Wheat Bread
One loaf for dinner, one to freeze

2 C fresh or frozen berries
1 C sugar, divided
1 1/2 C white flour
1 C whole wheat flour
1 t baking powder
1 t baking soda

1/2 t cinnamon
1/4 t salt
5 eggs
1/2 C + 1 T oil
1 t vanilla
1 C chopped walnuts

Sprinkle washed berries with 1/2 cup sugar. Combine dry ingredients and set aside. Beat eggs and beat in remaining 1/2 C sugar until mixture thickens. Stir in oil and vanilla. Add walnuts and flour and stir just until moistened. Fold in blueberries and any juice that may have accumulated; do not overmix. Pour mixture into two, greased loaf pans and bake at 350 degrees, 50-60 minutes. Remove from pans and cool before slicing.

84 Blueberry Pineapple Bread
These two loaves freeze well

3 C flour
2 t baking powder
1 t soda
1/2 t salt
1 1/2 t lemon juice
1 C chopped nuts

2/3 C shortening
1 1/3 C sugar
4 eggs
1 C crushed pineapple, drained
1/2 C flaked coconut
2 C fresh berries

Combine dry ingredients. Cream shortening and sugar until smooth; add eggs, milk, lemon juice and pineapple. Add dry ingredients and do not overbeat. Fold in berries, nuts and coconut. Pour batter into 2 greased and floured bread pans. Bake at 350 degrees, 35-40 minutes.

85   Blueberry Crunch Coffee Cake
Try this for brunch

Cake:
1/4 C butter
1/2 C sugar
1 egg
1 C flour
1t baking powder
1/4 t salt
1/3 C milk
1/2 t vanilla
2 C fresh or frozen berries

Topping:
1/4 C butter
1/2 C sugar
1/3 C flour
1/2 t cinnamon
1/4 t nutmeg

Make cake batter by creaming butter, sugar and egg; add dry ingredients alternately with milk and vanilla. Spread batter in a greased 9X9 square baking pan and top with berries. Combine the topping ingredients using a fork or pastry blender. Sprinkle mixture over berries. Bake at 350 degrees, 25-30 minutes or until it tests done.

86   Blueberry Brunch Cake
Poppyseeds and a touch of lemon

Cake:
1/2 C butter
1/2 C sugar
2 t grated lemon peel
1 egg
1 1/2 C flour
2 T poppy seed
1/t baking powder
1/4 t salt
1/2 C sour cream

Filling:
2 C blueberries
1/3 C sugar
2 t flour
1/4 t cinnamon

Glaze:
1/3 C powdered sugar
1-2 t milk

Cream butter and sugar; add eggs and lemon peel. Combine dry ingredients and add to creamed mixture with sour cream. Spread batter in a greased 9 inch springform pan, spreading batter one inch up sides of pan to hold filling. Combine filling ingredients and spread over batter. Bake at 350 degrees, 40-50 minutes. Combine glaze ingredients and drizzle over cake.

87 Blueberry Lemon Bread
The lemon tang is perfect

1 C sugar
1/2 C butter
2 eggs
2 C flour
1 1/2 t baking powder

1/4 t salt
1 C blueberries
2 t grated lemon peel
1/4 C lemon juice
1/3 C sugar

Cream sugar and butter; beat in eggs one at a time. Combine dry ingredients
and stir into creamed mixture. Fold in berries and lemon peel. Pour batter into a greased loaf pan. Bake at 350 degrees, 50-60 minutes or just until done. Combine lemon juice and sugar and pour over hot loaf. Cool in pan for 10 minutes.

88 Banana Blueberry Bread
A very old recipe

2 C flour
1 t baking soda
1/2 t salt
1/2 t cinnamon
1 C butter
3/4 C sugar

2 large eggs
1 C mashed banana
1/2 C sour cream
1 C fresh blueberries
1/2 C chopped nuts

Combine dry ingredients. Cream butter and sugar; add eggs, bananas, and sour cream. Add dry ingredients and beat just until blended. Fold in nuts and berries. Bake at 350 degrees, 1 hour or until done. Cool in pan.

89 Blueberry Walnut Cake

A three-layer masterpiece with lemon filling

1 1/2 C whipping cream
2 t vanilla
3 eggs
1 1/2 C flour
1 1/2 C sugar
1 C ground walnuts
2 t baking powder
1/4 t salt
1 C fresh blueberries

Lemon filling:

1/2 C margarine
1 1/2 C powdered sugar
1/2 t vanilla
1 T lemon juice
2 egg yolks

Whip cream until stiff; add vanilla. Beat eggs until light and lemon colored; fold into cream. Combine remaining cake ingredients except berries and fold gently into cream. Fold in berries. Pour into three, oiled 8 inch layer pans. Bake at 350 degress, 30-35 minutes. Filling: Beat sugar, margarine, juice and vanilla until very light; add yolks one at a time, beating until fluffy. Put half of filling between each layer; sprinkle top of cake with powdered sugar. Refrigerate until serving time.

90   Sour Cream Blueberry Cake

1 1/2 C flour
1/2 C sugar
1/2 C butter
1 1/2 t baking powder
1 egg
1 t vanilla

3 C fresh or frozen berries
2 C sour cream
2 egg yolks
1/2 C sugar
1 t vanilla

Combine first six ingredients and mix well. Spread mixture into a buttered 9 or 10 inch springform pan. Sprinkle evenly with berries. (If using frozen berries, wipe away ice crystals before adding.) Combine sour cream, egg yolks, sugar and vanilla. Spread over berries. Bake at 350 degrees, 1 hour.

91 Blueberry Pudding
A Southern favorite, with wine sauce

1 C sugar
1 egg
4 T corn oil
1 C flour
1/2 C buttermilk

4 C blueberries
Sauce:
2 T margarine
3/4 C sugar
1 C wine

Beat together sugar, egg, oil and buttermilk; add flour and mix
well. Fold in blueberries. Bake in an 8X8 inch greased square pan
at 350 degress, 30 minutes. To make sauce: Cream sugar and
margarine until very light; add scuppernong wine (typically
Southern) or other wine of choice.

92   **Blueberry Custard**
A delicious filling for cream puffs or meringues

1/2 C sugar
3 T corn starch
1/4 t salt
1/2 C water

1 T lemon juice
2 T butter
3 C blueberries
3 egg yolks

Combine dry ingredients in a saucepan; stir in water and lemon juice and blend well. Cook over medium heat until thickened. Remove from heat and stir in butter. Add egg yolks one at a time blending throughly. Add blueberries and cool. Top servings with sweetened whipped cream, and garnish with blueberries.

93   Blueberry Batter Cake
Quick, easy, and beautiful!

2 C blueberries
juice of 1/2 lemon
1/2 C sugar
3 T butter or margarine
1/2 C milk
1 C sifted flour

1 t baking powder
1/4 t salt
2/3 C sugar
1 T cornstarch
1/4 t salt (for topping)
1 C boiling water

Oil or spray an 8X8 inch baking pan; put cleaned berries into pan and sprinkle with lemon juice. In a mixing bowl, cream 1/2 C sugar and butter; add milk alternately with flour, baking powder and salt which have been mixed together. Pour batter over berries. Combine 2/3 C sugar, cornstarch and salt; sprinkle over top of cake.. Pour boiling water over all. Bake at 350 degrees, 45 minutes to 1 hour.

94 Ritzy Blueberry Dessert
A nutty meringue with topping

3 egg whites
1/2 t cream of tartar
1 C sugar
1 C finely crushed snack cracker crumbs
1/2 C chopped walnuts
1 4 oz pkg frozen whipped topping, thawed
1 can blueberry pie filling

Beat egg whites and cream of tartar until frothy; add sugar
gradually and beat until stiff. Fold in cracker crumbs and nuts.
Pile mixture into an ungreased 8X8 inch pan and bake at 350
degrees, 25 minutes. (If baking in glass, reduce oven temperature
to 325 degrees.) Cool. Cover meringue with whipped topping and
top with blueberry filling. Serves 9.

95 Blueberry Buckle

A classic!

2 C flour
2 1/2 t baking powder
1/4 t salt
1/2 C butter
1/2 C sugar
1 egg
1/2 C milk

Topping:
2 C fresh blueberries
1/2 C flour
1/4 C butter
1/2 C sugar
1/2 t cinnamon

Combine dry ingredients. Cream butter and sugar; add egg. Add dry ingredients alternately with milk. Spread batter in a greased 13X9 pan. Sprinkle with berries. Combine remaining topping ingredients until crumbly and sprinkle over berries. Bake at 350 degrees, 35-45 minutes. Serve warm. Serves 8-12.

96 Blueberry Crisp

4 C blueberries
2/3 C sugar
2 T flour
1/2 t lemon juice
1/4 t nutmeg
1/2 t cinnamon

Topping
1 C flour
1/4 C brown sugar
1/2 C oatmeal
1/2 C margarine

Combine berries with flour, sugar and spices. Put mixture in a greased 8X12 inch baking dish. Mix topping ingredients and sprinkle over berries. Bake at 350 degrees, 35-45 minutes or until bubbly. Serve warm with ice cream.

97   Blueberry Pie
Who doesn't name this as a favorite?

4 C fresh or frozen berries
5 T flour
3/4 C sugar
1/2 t cinnamon
1 T lemon juice
1 T butter
Pastry for a two crust pie

Combine all ingredients except butter and pile into 9-10 in pie shell. Dot with butter. Cover with top crust in which slits have been cut. Sprinkle with white sugar. Bake at 400 degrees, 40 minutes or until bubbling and nicely brown.

98   **Fresh Blueberry Glace Pie**
Natural sweetness of the berries is almost enough

1 9 inch baked pieshell

5 C blueberries

1/2 C sugar

3 T cornstarch

1/4 t salt

1/4 C water

1 T lemon juice

whipped cream or whipped topping

Crush 1 1/2 C berries. Combine sugar, salt, cornstarch and water; stir into crushed berries and cook over medium heat, stirring constantly until thickened. Add lemon juice and cool. Fold remaining fresh berries into cooked mixture until all are coated. Turn mixture into pieshell. Top with whipped cream or topping and chill.

99 Blueberry Cream Puff Pie
An old family favorite

Cream Puff Pastry:

1/2 C water	1/8t salt
1/4 C butter	2 eggs
1/2 C flour	

Bring water and butter to a boil in a saucepan; add flour and salt and stir until mixture leaves the side of pan. Remove from heat and beat in eggs one at a time. Beat until mixture is smooth and velvety. Spread in just the bottom of a greased glass 9 inch pie plate. Bake at 400 degrees, 50 minutes.
(Mixture stays flat in center and puffs into a big ring around plate.

Custard Filling:

1/2 C sugar	2 C milk
1/2 t salt	1 egg yolk
2 T corn starch	1/2 C whipped cream or topping
1 T flour	1 t vanilla
2 C unsweetened fresh blueberries	

Combine sugar, salt, cornstarch and flour in a saucepan; stir in milk and cook for one minute, stirring constantly. Stir some of hot mixture into egg yolk and add back to sauce pan. Cook gently for one minute. Cool. Stir cream and vanilla into custard. Put cold custard into pastry and top with berries. Serves 8.

100 Peach and Blueberry Pie
Two luscious summer fruits together

Pastry for a two crust pie
3 C sliced fresh peaches
1 C fresh blueberries
2 T lemon juice

1 C sugar
2 T tapioca
1/2 salt
2 T butter

Sprinkle lemon juice over fruits; stir in sugar, salt and tapioca. Pour mixture into pieshell and dot with butter. Put on top crust and sprinkle with white sugar. Bake at 375-400 degrees, 40-50 minutes.

101 Lemon Blueberry Cheese Pie

I baked 9 inch pie shell
Filling:
1 8 oz pkg cream cheese
1 can sweetened condensed milk
1/2 C lemon juice
1 t vanilla
1 C fresh or frozen berries

Glaze:
1 C sugar
2 T corn starch
1 C water
2 C fresh or frozen berries

Beat softened cream cheese until fluffy; beat in milk, lemon juice
and vanilla. Fold in berries and pour into pie shell. To make
glaze: Crush 1/2 C berries and combine in a sauce pan with water,
sugar and cornstarch. Cook and stir until mixture is thick and
clear, about two minutes. Put mixture through a strainer and cool.
Arrange remaining 1 1/2 C berries over cheese filling and pour
glaze over berries. Chill. Serves 8.

102 Polka Dot Pie
So pretty !

1 3 oz pkg lemon gelatin
2/3 C boiling water
1/2 C cold water
ice cubes

3 1/2 C frozen whipped topping
2 C fresh blueberries
1 graham cracker crust

Dissolve gelatin in boiling water. Combine water and ice cubes to make 1 1/4 C. Add to gelatin and stir until no more ice melts; remove ice. Blend whipped topping into gelatin with a wire whisk. Fold in 1 /2 C blueberries and pour into crust. Chill. Garnish with more topping, if desired, and with remaining berries. Serves 6.

103 Quickie Blueberry Pie
So easy. No crust to roll

1/2 C butter
1 T sugar
1 C flour
1 can blueberry filling

Topping:
1 egg
1/2 C sugar
1/4 C flour
1/2 C milk

Melt butter and add 1 T sugar. Stir into 1 C flour and stir until mixture forms a ball. Press into bottom and sides of a 9 inch pie pan. Pour filling into crust. To make topping: Beat egg with 1/2 C sugar; blend in flour and milk until smooth. Put topping over filling. Bake at 350 degrees, 50-60 minutes.

104 Barb's Blueberry Sky Pie
Makes two for a holiday dinner

1 8 oz pkg cream cheese
1 6 oz can frozen lemonade
1 can sweetened condensed milk

1 8 oz carton whipped topping
1 can blueberry pie filling
2 graham cracker crusts

Warm cream cheese to room temperature and beat until creamy;
beat in lemonade, then milk. Fold 3/4 of the whipped topping and
2/3 of the pie filling into mixture, reserving the rest for topping.
Pour mixture into crusts and chill until set. Pile remaining
whipped topping in center of each pie and make a well in the
center. Fill center with remaining pie filling.

VERY
CRANBERRY

105 Cranberry Tea
A wonderful winter hot drink!

1 bag cranberries
3 qts water
2 cinnamon sticks

1/2 C orange juice
1/3 C lemon juice

Bring cleaned berries to a boil with water and cinnamon sticks; boil until berry skins pop. Remove cinnamon sticks and push mixture through a strainer. Add juices, sugar and more water to taste. Serve hot.

106 Hot Cranberry Cider

1 1/2 qts cranberry juice cocktail
1 1/2 qts apple cider
Rum to taste (optional)
Candy cinnamon or peppermint sticks
Heat juices until hot. Serve in mugs with candy sticks. Serves 8.

107 Cranberry Apricot Cooler

1 pt cranberry cocktail   12 oz apricot nectar
2 T lemon juice

Combine juices, ice and serve. Serves 4.

108 Cranberry Relish
Is there Thanksgiving Dinner without it?

1 pkg cranberries   2 unpeeled apples
2 unpeeled oranges   2 C sugar

Remove apple cores, and orange seeds. Grind all fruits finely and
add sugar. Chill.

109 Cooked Cranberry Salad

1 3 oz pkg cherry gelatin
2 C raw cranberries
1 C sugar
2 C water

1 C apple, chopped
1 C celery, chopped
1/2 C nuts, chopped

Cook cranberries and sugar in 1 1/2 C water until skins pop. Dissolve gelatin in remaining 1/2 cup water (heated to boiling.) Stir gelatin, apple, celery and nuts into berries. Chill until set.

110 Cranberry Mashmallow Salad

2 C raw cranberries
4 C miniature marshmallows
1/2 C sugar

1/4 C chopped apple
1/2 C chopped nuts
1 C whipping cream,

Grind cranberries or process with a food processor; stir in marshmallows and sugar and let chill overnight. Whip cream and fold into mixture with nuts and apple. Pile each serving on a pineapple slice. Serves 12.

111 Molded Cranberry Salad

1 C ground cranberries
1 C ground apple
1 C sugar
1 3 oz pkg lemon gelatin

1 C hot water
1 C pineapple syrup
1/2 C halved red grapes
1/2 C chopped nuts

Combine canberries, apple and sugar. Dissolve gelatin in hot water; add syrup (drained from canned pineapple). Chill until partially set; add fruit mixture, grapes and nuts. Chill until firm.

112 Cranberry Butter

1 C raw cranberries
1 1/2 C powdered sugar
1 C butter

1 T lemon juice
1 t ground ginger

Blend cranberries in a blender until finely chopped. In a mixer bowl, beat butter and sugar light and fluffy. With mixer running, blend in berries, juice and ginger until mixture is smooth. Chill. Serve with pancakes, waffles, or French toast. Makes 2 cups.

113 Cranberry Nut Bread
So pretty

2 C flour
1 C sugar
1 1/2 t baking powder
1/2 t soda
1 t salt
1 egg

3/4 C orange juice
1 T grated lemon rind
3 T oil
1/2 C chopped nuts
2 C fresh or frozen berries

Combine dry ingredients in a mixing bowl. Combine egg, juice, rind and oil; add all at once to dry ingredients and mix just until blended. Fold in nuts and cranberries. Pour batter into a greased bread pan. Bake at 350 degrees, 1 hour. Cool before slicing.

114 Cranberry Cookies
A big batch of little jewels

3 C flour
1 t baking powder
1/2 t salt
1/4 t soda
1/2 C shortening
1 C sugar

3/4 C brown sugar, packed
1/4 C milk
2 T orange juice
1 egg
1 C chopped nuts
2 1/2 C cranberries

Cream shortening, and sugars until blended. Add milk, juice and egg; blend in dry ingredients. Stir in coarsely chopped berries and nuts. Drop on greased baking sheets and Bake at 375 degrees, 12-15 minutes. Cookies may be sprinkled with sugar while hot. 5-7 dozen cookies.

115 Cranberry Pound Cake
Easy, fast and delicious!

1 pkg yellow cake mix   1 C cranberries
1 C plain yogurt   1/2 C chopped nuts
4 eggs

Combine cake mix with yogurt and eggs; blend until moistened,
then beat for two minutes. Fold in chopped cranberries and nuts.
Pour batter into a greased and floured bundt pan, or tube pan.
Bake at 350 degrees, 35-45 minutes. Cool for 15 minutes and
remove from pan.

116 Baked Cranberry Pudding

1 1/2 C flour
2 t baking powder
1/2 t salt
2 T shortening

1/2 C sugar
1 egg
1/2 C milk
1 1/2 C cranberries

Cream sugar and shortening; add egg, milk and combined dry ingredients. Fold in whole cranberries. Pour batter in a 8X8 inch pan. Bake at 350 degrees, 30 minutes. Serve with Sauce.

Pudding Sauce
1 C powdered sugar
2 T butter
1 egg

Cream shortening and sugar; add egg and beat to a rich creamy consistency.

117 Cranberry Nut Pie
Makes its own crust

2 C fresh or frozen berries
1/2 C chopped nuts
1/2 C brown sugar
2 eggs

1 C flour
1 C sugar
1/2 C melted butter
1 t grated orange rind

Grease a 10 inch pie plate. Spread berries and nuts in plate and sprinkle with sugar. Beat eggs and add remaining ingredients; beat until smooth. Pour over berry-nut mixture. Bake at 325 degrees 1 1/4 hours or until top is a light golden brown. (Do not underbake.) Serve warm with vanilla ice cream.

The Grand Cookbook Series

**The Big Fat Red Juicy
Apple Cook Book**

**A Very Berry
Cookbook**

Cherry Time

Tasty Taters

Fish Food

Say Cheese

Do you need a hostess or birthday gift for a friend? Order extra *Grand Cookbooks* now to have on hand or let us send your gift for you.

L.E.B. Inc.
27599 Schoolcraft
Livonia, Michigan 48150

❑ The Big Fat Red Juicy Apple Cook Book
❑ Cherry Time
❑ Tasty Taters
❑ Say Cheese

❑ Fish Food
❑ A Very Berry Cookbook
❑ Merry Cookie

Please send the cookbooks checked above at $6.95 plus $1.00 postage and handling per book to:

(Please Print)

Name _____

Address _____

City_____State_____Zip _____

Enclose a gift card from:

_____

(Your Name)

❑ The Big Fat Red Juicy Apple Cook Book
❑ Cherry Time
❑ Tasty Taters
❑ Say Cheese

❑ Fish Food
❑ A Very Berry Cookbook
❑ Merry Cookie

Please send the cookbooks checked above at $6.95 plus $1.00 postage and handling per book to:

(Please Print)

Name _____

Address _____

City_____State_____Zip _____

Enclose a gift card from:

_____

(Your Name)